INSPIRATIONAL QUOTES COLORING BOOK

CRYSTAL
COLORING BOOKS

Copyright © 2018 Crystal Coloring Books
All rights reserved.

ISBN-13: 978-1986302708
ISBN-10: 1986302709

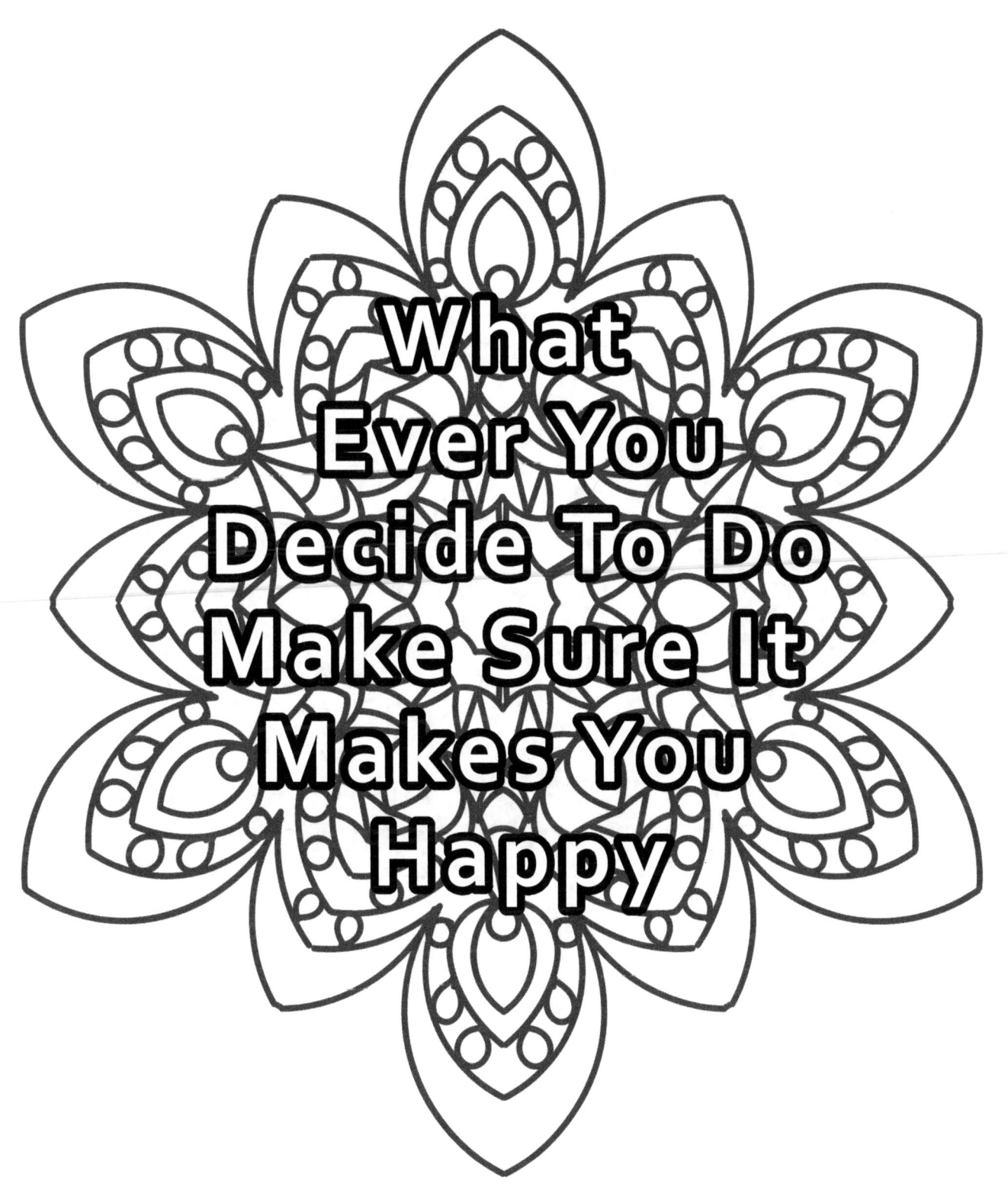

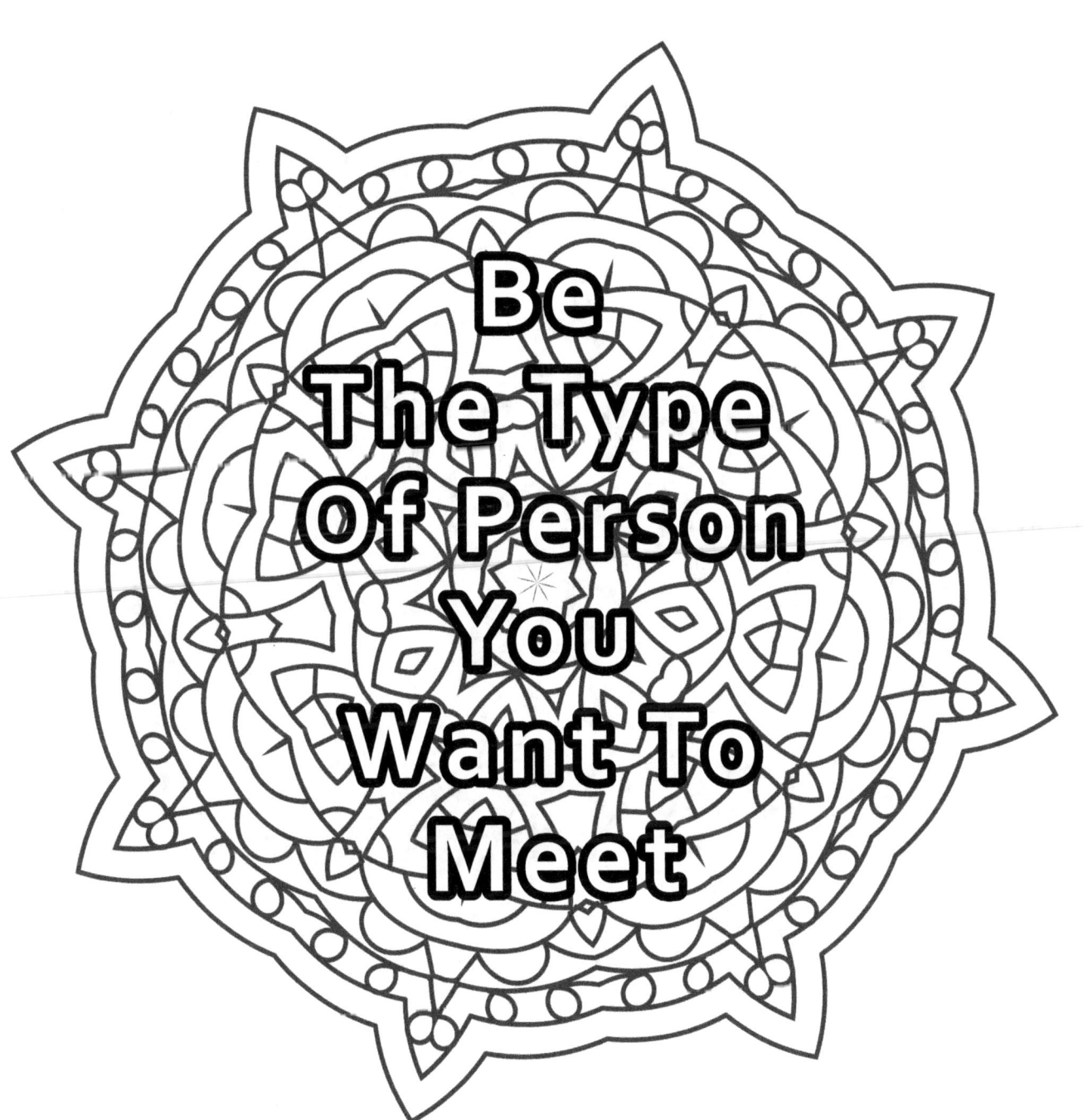

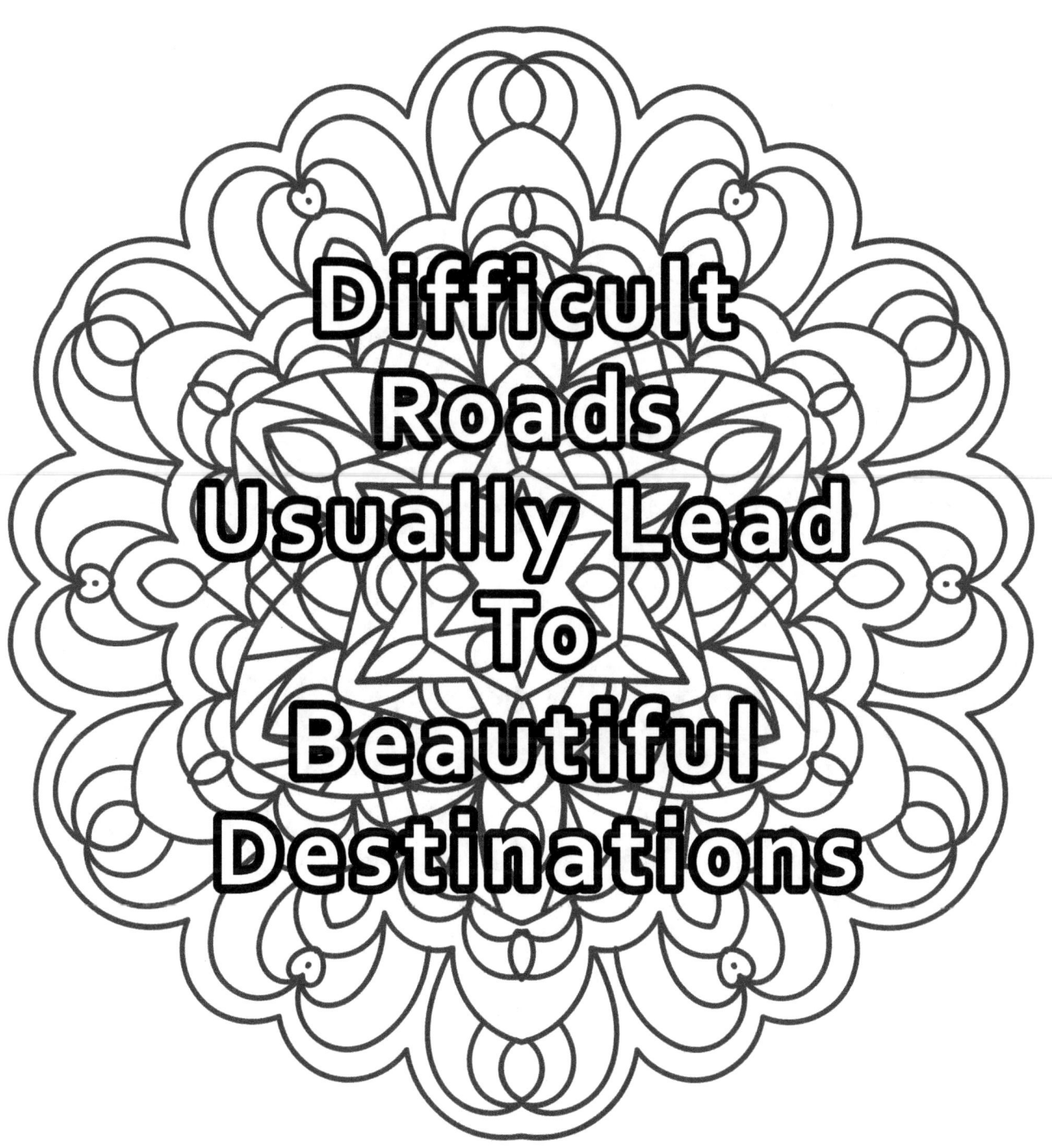

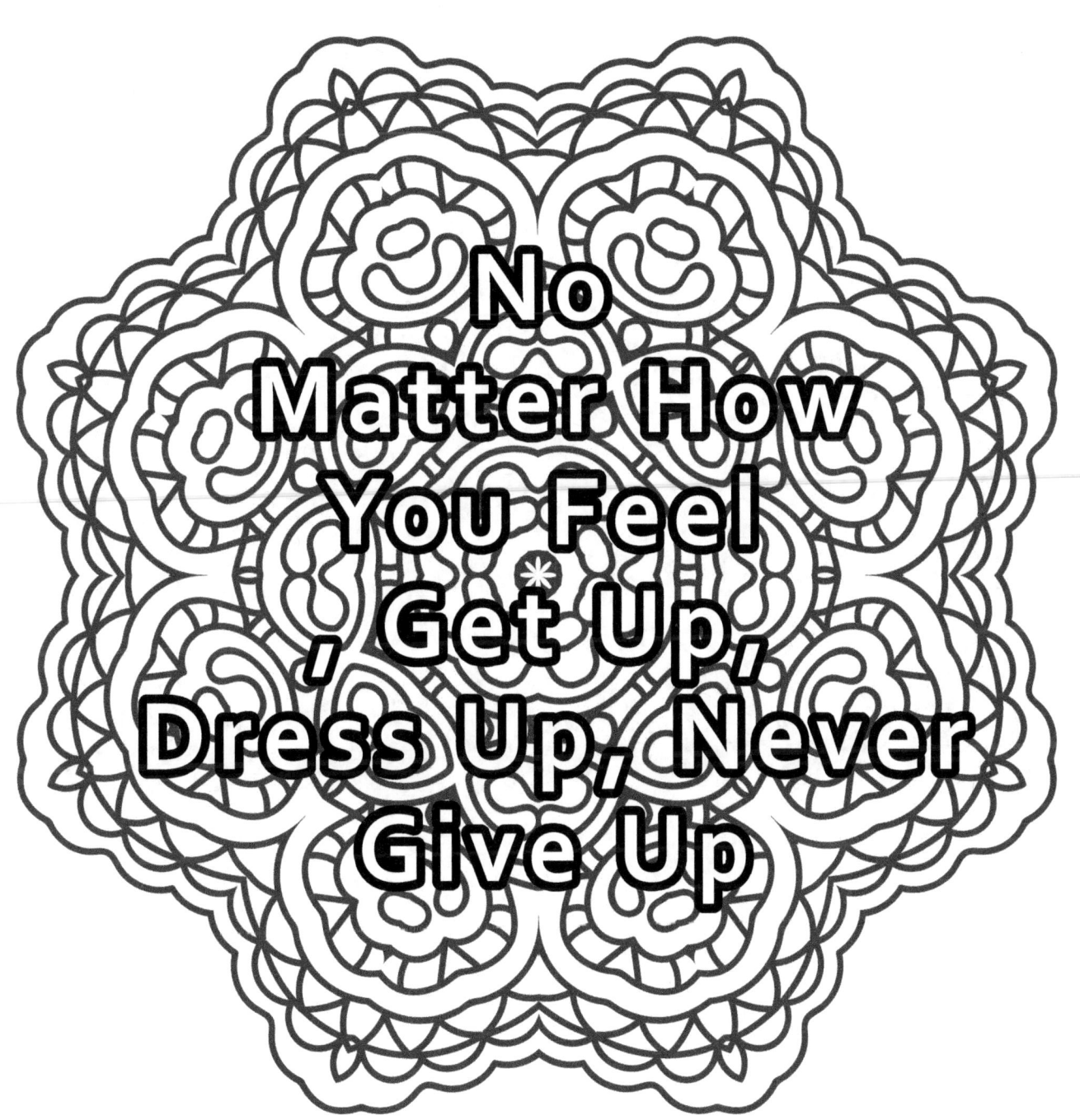

COLOR TEST PAGE

www.ingramcontent.com/pod-product-compliance
Lightning Source LLC
Chambersburg PA
CBHW062127220526
45471CB00010B/3907